The Starlit Mind

*A Cosmic Guide to Nurturing
Your Neural Universe*

Darrell Sheppeard

CONTENTS

"The Starlit Mind: A Cosmic Guide to Nurturing Your Neural Universe"

PREFACE:

The Cosmic Symphony of Self-Discovery

I n the vast expanse of the universe, a symphony of celestial wonders unfolds before our eyes, each note a testament to the boundless potential that lies within the cosmic tapestry of our existence. As we journey through the stars, we are drawn to the shimmering constellations that adorn the night sky, their radiant energy a beacon of hope and inspiration, guiding us on our voyage through the cosmos of the mind.

This book is an invitation to embark on a grand odyssey of self-discovery, a celestial voyage that will take you on a journey through the uncharted realms of your inner universe, unlocking the boundless potential that resides within the labyrinthine depths of your mind. As you explore the celestial realms of neurogenesis and synaptogenesis, you will uncover the secrets of the mind-body connection, delve into the mysteries of nutrition and brain health, and venture into the realms of exercise, sleep, mindfulness, meditation, and cognitive training.

Each chapter of this cosmic symphony will reveal a new facet of the intricate tapestry of your mind, weaving together explanations and metaphorical examples that will illuminate the path to personal growth, self-discovery, and boundless potential. As you immerse yourself in the celestial melodies of this grand opus, you will learn how to harness the power of neurogenesis and synaptogenesis, transforming your brain into a fertile landscape of neural pathways and synaptic connections that will empower you to transcend your limitations and reach for the stars.

But this is more than just a book; it is a cosmic journey, an odyssey through the stars that will awaken the explorer within you, igniting the spirit of adventure, curiosity, and wonder that lies dormant within the depths of your soul. As you journey through the celestial realms of your inner universe, you will uncover the secrets of your own potential, unlocking the doors to a world of infinite possibility and boundless creativity.

As you stand on the precipice of your own cosmic adventure, we invite you to join us on this celestial voyage, to embrace the spirit of discovery that has guided humankind throughout the ages, and to chart your own course among the stars. Together, we will explore the vast expanse of our inner cosmos, uncovering the secrets of our minds and unlocking the boundless potential that lies within our hearts and souls.

So, dear traveler, as you embark on this grand odyssey of self-discovery, let the celestial symphony of your dreams resonate throughout the cosmos, a testament to the boundless potential of the human spirit and the unquenchable thirst for growth, learning, and self-discovery that lies at the heart of our existence.

Let the cosmic journey begin.

CHAPTER 1: A CELESTIAL DAWN: THE BIRTH OF NEW NEURAL PATHWAYS

The human mind is a magnificent universe, teeming with boundless potential and infinite possibilities. At the heart of this celestial expanse lie the wondrous processes of neurogenesis and synaptogenesis, the birth of new neurons and synaptic connections that form the very fabric of our cognitive cosmos. Like the radiant stars that illuminate the night sky, these processes light the way toward a brighter, more brilliant future, one filled with self-discovery, growth, and personal fulfillment.

THE GENESIS OF NEURONS AND SYNAPSES: UNVEILING THE COSMIC DANCE

Just as a newborn star ignites the vast expanse of the cosmos, neurogenesis represents the birth of new neurons within our brains, a process that plays a crucial role in learning and memory. In this celestial dance, neurons are akin to stars, each one bursting forth into existence, adding its unique brilliance to the tapestry of our cognitive universe.

Synaptogenesis, on the other hand, is the formation of

new connections between neurons, much like the intricate constellations that adorn our night sky. Each synapse represents a bridge between stars, a thread of cosmic energy that binds the celestial tapestry together, weaving an intricate web of neural pathways that shape our thoughts, emotions, and actions.The Power of Brain Plasticity: A Universe in Constant Flux

The universe is a realm of constant change, a dynamic cosmos that is forever evolving, expanding, and transforming. Our minds, too, are subject to this cosmic principle, thanks to the incredible power of brain plasticity. This innate ability of our brains to adapt, learn, and change over time allows us to reshape our neural landscape, forging new connections and pathways as we journey through the vast expanse of our cognitive universe.

ACHIEVING YOUR DREAMS: HARNESSING THE POWER OF THE COSMOS WITHIN

As we embark on our celestial voyage of self-discovery, we must learn to harness the power of neurogenesis and synaptogenesis, using these processes to reshape our cognitive universe and unlock the boundless potential that lies within our minds. By forging new neural pathways and synaptic connections, we can transcend the limitations of our past, charting a new course toward a brighter, more brilliant future, one filled with growth, learning, and personal fulfillment.

Much like the celestial bodies that populate the night sky, our dreams represent the guiding stars of our existence, radiant beacons of hope and inspiration that guide us on our journey through the cosmos of the mind. By harnessing the power of neurogenesis and synaptogenesis, we can reshape our cognitive universe, forging new connections and pathways that will empower us to reach for the stars and achieve our dreams.

In this celestial odyssey, we will explore the myriad ways in which we can unlock the boundless potential of our minds, embracing the power of brain plasticity to transform our lives, and charting a new course towards a brighter, more brilliant future, one filled with self-discovery, growth, and personal fulfillment.

CHAPTER 2: THE SYMPHONY OF THE MIND: NEUROGENESIS AND SYNAPTOGENESIS IN HARMONY

Picture your brain as a grand symphony orchestra, with each musician representing a neuron. In this extraordinary ensemble, synapses are the notes played, connecting the musicians, and creating a harmonious, ever-evolving melody of thought, emotion, and action. The continuous growth and connection of these musicians and their notes are what neurogenesis and synaptogenesis represent, ensuring the orchestra's repertoire remains diverse and its performance unparalleled.

THE MUSICIANS AND THEIR INSTRUMENTS: NEURONS AND SYNAPSES

The role of neurons and synapses in brain function is akin to musicians and their instruments in an orchestra. Each neuron, like a musician, has a specific role, sending electrical and chemical signals through a complex network of connections. The synapses, similar to the notes, are the points of contact between neurons, where signals are transmitted, and communication is established. This elaborate symphony of signals and connections enables us to think, feel, learn, and adapt to our ever-changing environment.

THE NEWCOMERS AND VIRTUOSOS: NEUROGENESIS AND LEARNING

Imagine a new musician joining the orchestra, bringing a fresh perspective, and enhancing the ensemble's performance. This is the essence of neurogenesis: the process of creating new neurons that enrich the brain's capabilities. Neurogenesis is especially crucial in regions of the brain associated with learning and memory, such as the hippocampus. As new neurons integrate into the existing network, they contribute to the formation of new memories and the enhancement of cognitive abilities, like virtuosos joining the orchestra and expanding its repertoire.The Evolving Melodies: Synaptogenesis and Neural Connections

In an orchestra, musicians must continuously adapt and form new connections to create a symphony that captivates its audience. Similarly, the process of synaptogenesis allows our brains to adapt and evolve by forming new neural connections. As we learn and experience, the brain alters its connections to accommodate and consolidate new information, much like an orchestra refining its melody. This ongoing process of synaptogenesis creates a dynamic, ever-changing network of connections that enables us to navigate the complexities of our lives.

THE CONDUCTOR'S INFLUENCE: FACTORS AFFECTING NEUROGENESIS AND SYNAPTOGENESIS

Just as a conductor's skill and vision can profoundly impact an orchestra's performance, certain factors influence the process of neurogenesis and synaptogenesis in our brains. Lifestyle choices, such as exercise, nutrition, and stress management, play a significant role in promoting or hindering the growth and connection of neurons. By understanding these factors and consciously making choices that foster neurogenesis and synaptogenesis, we can become the conductors of our brain's symphony, guiding it toward greater harmony and success.

As we explore the science behind neurogenesis and synaptogenesis, we come to appreciate the remarkable symphony of our minds. By acknowledging the roles of neurons and synapses, understanding the processes of neurogenesis and synaptogenesis, and identifying the factors that influence these processes, we can begin to take control of our brain's orchestra and compose the masterpiece of our lives.

In the chapters to follow, we will delve deeper into the strategies and techniques that can help you optimize your lifestyle, enhance your brain health, and foster the growth and connection of neurons. By doing so, you will be well on your way to harnessing the power of neurogenesis and synaptogenesis, unlocking your full potential, and, ultimately, living the life of your dreams.

Together, let us embark on this journey of self-discovery and transformation, mastering the art of conducting the grand symphony of our minds and creating harmonious, ever-evolving melodies that resonate with our aspirations and desires. Embrace the beauty and complexity of your brain's orchestra and let its captivating performance guide you toward a future filled with infinite possibilities and boundless potential.

CHAPTER 3: THE MIND-BODY DANCE: A HARMONIOUS PARTNERSHIP

Picture the mind and body as two dancers engaged in a graceful, intricate dance, intimately connected and interdependent. Their movements are synchronized, reflecting the delicate balance and harmony between thoughts, emotions, and brain function. In this chapter, we will explore the captivating dance between the mind and body, uncovering the profound impact that our thoughts and emotions can have on our brain's health and well-being.

THE CHOREOGRAPHY OF THOUGHTS AND EMOTIONS: THEIR INFLUENCE ON BRAIN FUNCTION

Thoughts and emotions are like the steps of the dance, intricately woven together to form the choreography that guides the dancers' movements. Our thoughts and emotions are deeply interconnected, with each influencing and shaping the other. This interplay affects not only our mental state but also our brain function, as our mind and body respond to the ever-changing rhythm of our thoughts and emotions.

Neurotransmitters, the chemical messengers of the brain, act as the music to which the dancers move, transmitting signals that influence our thoughts, emotions, and actions. Positive thoughts and emotions can foster the release of neurotransmitters, such as serotonin and dopamine, that promote feelings of happiness and well-being. Conversely, negative thoughts and emotions can trigger the release of stress hormones, such as cortisol, leading to a cascade of physiological responses that can affect both our mental and physical health.

THE STUMBLING STEPS: STRESS AND NEGATIVE EMOTIONS

In the dance of the mind and body, stress and negative emotions are like stumbling steps, disrupting the harmony and balance between the dancers. Chronic stress and negative emotions can hinder neurogenesis and synaptogenesis, impairing the brain's ability to create new neurons and form new connections. This disruption can have lasting effects on our cognitive abilities, memory, and emotional well-being.

Additionally, prolonged exposure to stress and negative emotions can exacerbate existing health conditions and increase the risk of developing new ones, such as anxiety, depression, and cardiovascular disease. In essence, the stumbling steps of stress and negative emotions can throw the entire dance off balance, affecting both the mind and body in profound ways.The Graceful Movements: Positive Thinking and Visualization

In contrast to the stumbling steps of stress and negative emotions, positive thinking and visualization are like graceful, fluid movements that elevate the dance of the mind and body to new heights. These practices can have a profound impact on brain health, enhancing neurogenesis, synaptogenesis, and overall cognitive function.

Positive thinking and visualization can stimulate the release of neurotransmitters associated with happiness and well-being, as well as promote the growth of new neurons and the formation of new connections. Furthermore, these practices can help to strengthen the mind-body connection, enhancing our ability to cope with stress, overcome challenges, and achieve our goals.

By consciously cultivating positive thoughts and engaging in visualization exercises, we can fine-tune the choreography of our mind-body dance, ensuring that our movements are graceful, harmonious, and in sync with our aspirations and desires.

As we delve deeper into the mind-body connection, we gain a deeper appreciation for the intricate dance between our thoughts, emotions, and brain function. By recognizing the impact of stress and negative emotions on our brain health and embracing the benefits of positive
thinking and visualization, we can take charge of our mind-body dance, guiding it toward greater harmony, balance, and well-being.

In the following chapters, we will explore practical techniques and strategies to help you nurture a positive mindset, reduce stress, and harness the power of visualization to improve your brain health and overall well-being. By applying these strategies, you will be well-equipped to steer the dance of your mind and body with grace and purpose, fostering a stronger, more resilient connection between the two.

Together, let us embark on this transformative journey, mastering the art of mind-body dance and gracefully navigating the complexities of life. As we become more attuned to the delicate balance between our thoughts, emotions, and brain function, we can create a harmonious, resilient partnership between our minds and bodies, allowing us to achieve our goals, overcome challenges, and live our lives to the fullest.

Embrace the captivating dance of your mind and body, and let its fluid, graceful movements guide you toward a future filled with health, happiness, and boundless potential. The dance floor awaits; it's time to take the first step toward a harmonious and thriving mind-body connection.

CHAPTER 4: NOURISHING THE MIND'S GARDEN: THE POWER OF NUTRITION FOR BRAIN HEALTH

P icture your brain as a lush, vibrant garden teeming with life and vitality. The neurons and synapses that make up this garden require essential nutrients to grow and flourish, much like plants require sunlight, water, and fertile soil. In this chapter, we will explore the vital role that nutrition plays in nurturing the garden of our minds, discovering the key nutrients and foods that support brain health, and learning how to create a brain-healthy meal plan.

THE FERTILE SOIL: DIET AND ITS IMPACT ON NEUROGENESIS AND SYNAPTOGENESIS

Just as the quality of soil is crucial for a thriving garden, the role of diet in promoting neurogenesis and synaptogenesis is paramount for a healthy brain. Consuming a nutrient-rich diet can provide the brain with essential building blocks, enabling it to create new neurons, form new connections, and maintain optimal function.

Conversely, a diet lacking essential nutrients can impair the brain's ability to grow and adapt, hindering neurogenesis and synaptogenesis. By nourishing our minds with a balanced, nutrient-dense diet, we can provide the fertile soil needed for our brain's garden to flourish and reach its full potential.

THE SUNLIGHT AND WATER: KEY NUTRIENTS AND FOODS FOR BRAIN HEALTH

Just as plants require sunlight and water to grow, our brain's garden depends on specific nutrients to thrive. Some key nutrients and foods that support brain health include:

Omega-3 fatty acids: These essential fats, found in foods such as fatty fish, walnuts, and chia seeds, play a crucial role in maintaining brain cell membrane integrity and promoting neuron communication.

Antioxidants: Found in colorful fruits and vegetables, antioxidants protect our brain from oxidative stress and inflammation, supporting overall brain health.

B vitamins: These essential nutrients, abundant in whole grains, legumes, and leafy greens, contribute to energy production and the synthesis of neurotransmitters.

Vitamin E: Found in nuts, seeds, and vegetable oils, vitamin E acts as a powerful antioxidant, protecting brain cells from damage and promoting cognitive function.

Choline: This essential nutrient, found in eggs, soybeans, and the liver, plays a vital role in memory and learning.

By incorporating these nutrient-rich foods into our diets, we can provide the sunlight and water needed for our brain's garden to thrive and prosper.

CULTIVATING A BRAIN-HEALTHY

MEAL PLAN: TIPS AND TRICKS

Creating a brain-healthy meal plan is like designing a well-tended garden, ensuring that each plant receives the nutrients, sunlight, and water it needs to flourish. Here are some tips for cultivating a nourishing and delicious meal plan to support your brain health:

Prioritize whole, unprocessed foods: Focus on nutrient-dense, minimally processed foods such as fruits, vegetables, whole grains, lean proteins, and healthy fats.

Eat a colorful array of fruits and vegetables: Aim to include a variety of colorful fruits and vegetables in your diet, as each color represents different antioxidants and phytonutrients that support brain health.

Incorporate healthy fats: Include sources of omega-3 fatty acids, such as fatty fish, walnuts, and chia seeds, to support brain function and neuron communication.

Stay hydrated: Ensure you're drinking enough water throughout the day, as dehydration can impair cognitive function and memory.

Practice mindful eating: Pay attention to your hunger and fullness cues, savor each bite, and enjoy your meals in a relaxed environment. Mindful eating can help you tune into your body's needs and foster a healthy relationship with food.

Plan ahead: Create a weekly meal plan and shopping list to ensure you have all the necessary ingredients on hand for nutritious, brain-boosting meals.

Experiment with new recipes: Keep your brain-healthy meal plan exciting and diverse by trying new recipes and discovering new foods that support brain health.

Moderation is key: While it's essential to prioritize nutrient-dense foods, it's also important to enjoy occasional treats in moderation. Balance and variety are crucial components of a sustainable, brain-healthy meal plan.

By cultivating a brain-healthy meal plan, you will be nurturing the garden of your mind with the essential nutrients, sunlight, and water it needs to grow and thrive. As you nourish your brain with wholesome, nutrient-rich foods, you'll begin to experience the profound impact that proper nutrition can have on your cognitive function, memory, and overall well-being.

Embrace the power of nutrition in nurturing the garden of your mind and let the vibrant, flourishing landscape of your brain guide you towards a future filled with clarity, focus, and boundless potential. Tend to your brain's garden with care, providing it with the essential nutrients and love it needs, and watch as it blossoms into a beautiful, thriving sanctuary of mental well-being.

CHAPTER 5: THE WINDS OF CHANGE: HARNESSING THE POWER OF EXERCISE FOR BRAIN HEALTH

I magine your brain as a vibrant, colorful kite soaring high in the sky, dancing gracefully in the winds of change. The winds of exercise provide the force and momentum needed for the kite to maintain its height and perform daring maneuvers, while the strings of neurogenesis and synaptogenesis keep the kite anchored and connected. In this chapter, we will explore the incredible benefits of exercise on brain health, delve into different types of exercise for optimal brain function, and learn how to create a personalized exercise plan.

THE GUSTS OF GROWTH: PHYSICAL ACTIVITY AND ITS IMPACT ON NEUROGENESIS AND SYNAPTOGENESIS

Physical activity acts as the gusts of wind that propel the brain's kite higher, stimulating neurogenesis and synaptogenesis in the process. When we engage in regular exercise, we not only improve our physical fitness but also promote the growth of new neurons and the formation of new connections within the brain. These new neurons and connections help enhance cognitive function, memory, and learning, keeping the kite of our brain soaring at its peak.

THE WINDS OF VARIETY: DIFFERENT TYPES OF EXERCISE FOR OPTIMAL BRAIN HEALTH

Just as varying wind patterns can challenge and strengthen a kite, incorporating different types of exercise into our routine can have a profound impact on our brain health. Here are some exercise modalities to consider:

Aerobic exercise: Activities such as running, swimming, and cycling increase blood flow and oxygen delivery to the brain, boosting neurogenesis, and improving cognitive function.

Strength training: Resistance exercises, like weightlifting and bodyweight exercises, can enhance synaptic plasticity and promote the release of growth factors that support brain health.

Flexibility and balance training: Practices such as yoga, Pilates, and tai chi can improve balance, coordination, and spatial awareness, all of which contribute to a healthy, agile brain.

Mind-body exercises: Integrative activities like meditation and mindfulness exercises can help reduce stress, improve emotional regulation, and enhance overall brain function.

By incorporating a diverse array of exercises into our routine, we can harness the winds of variety to keep our brain's kite soaring high and adapting to new challenges.

CHARTING YOUR FLIGHT PATH: CREATING A PERSONALIZED EXERCISE PLAN

Designing a personalized exercise plan is like charting the flight path of your brain's kite, ensuring it remains aloft and dances gracefully in the winds of change. Here are some tips for creating an effective exercise plan that supports optimal brain health:

Assess your current fitness level: Evaluate your current physical abilities, strengths, and weaknesses to help you tailor an exercise plan that meets your needs and goals.

Set achievable goals: Establish realistic, measurable goals that are specific to your desired outcomes, such as improving cognitive function, reducing stress, or enhancing memory.

Choose activities you enjoy: Select exercises that you genuinely enjoy and look forward to, as this will increase the likelihood of sticking to your plan.

Incorporate variety: Mix in different types of exercise to keep your routine fresh, engaging, and challenging for your brain.

Schedule your workouts: Designate specific days and times for your exercise sessions, treating them as non-negotiable appointments with yourself.

Monitor your progress: Track your workouts and evaluate your progress regularly, making adjustments as needed to ensure you continue to challenge your brain and body.

By charting your flight path and embracing the power of exercise, you will be well on your way to keeping your brain's kite soaring high, dancing gracefully in the winds of change, and maintaining its connection to the strings of neurogenesis and synaptogenesis. As you engage in regular physical activity, you will begin to experience the profound impact that exercise can have on your cognitive function, memory, and overall well-being.

Embrace the transformative power of exercise in propelling the kite of your mind to new heights and let the winds of physical activity guide you toward a future filled with clarity, focus, and boundless potential. Set your brain's kite soaring with the gusts of growth and variety provided by exercise, and watch as it dances gracefully in the sky, reflecting the vibrant, thriving landscape of your mental well-being.

CHAPTER 6: THE MOONLIT SANCTUARY: UNVEILING THE SECRETS OF SLEEP FOR BRAIN HEALTH

Envision your brain as a bustling city, illuminated by the sparkling lights of neurons firing and synapses connecting, orchestrating the daily symphony of life. At night, the city transitions to a moonlit sanctuary, where the quiet hum of activity continues but at a more subdued pace, allowing the city to regenerate and rejuvenate. In this chapter, we will unveil the secrets of sleep and its profound impact on brain health, discover tips for improving sleep quality, and learn how to create a sleep-friendly environment.

THE REGENERATIVE NIGHT: SLEEP'S ROLE IN NEUROGENESIS AND SYNAPTOGENESIS

Sleep serves as the regenerative night for our brain's bustling city, providing the essential rest and restoration needed for neurogenesis and synaptogenesis to occur. During sleep, our brain consolidates memories, forms new connections, and cleanses itself of toxins and waste products. These processes are crucial for maintaining optimal cognitive function, memory, and learning.

A lack of quality sleep can hinder neurogenesis and synaptogenesis, leaving our brain's city dimmed and less vibrant. Prioritizing restful, restorative sleep is essential for fostering a healthy, thriving brain and ensuring the city remains illuminated and bustling with life.

THE ART OF SLUMBER: TIPS FOR IMPROVING SLEEP QUALITY

Mastering the art of slumber is like learning to navigate the winding streets and alleys of our brain's city, ensuring that each corner receives the rejuvenating moonlight needed to thrive. Here are some tips for improving sleep quality and mastering the art of slumber:

Establish a consistent sleep schedule: Go to bed and wake up at the same time each day, even on weekends, to help regulate your body's internal clock.
Create a bedtime routine: Develop a relaxing pre-sleep routine to signal your brain that it's time to wind down and prepare for rest.
Limit exposure to screens: Reduce exposure to blue light from screens in the evening, as it can interfere with the production of the sleep hormone melatonin.
Watch your caffeine intake: Be mindful of your caffeine consumption, especially in the afternoon and evening, as it can disrupt your sleep.
Practice relaxation techniques: Incorporate relaxation exercises, such as deep breathing, meditation, or gentle stretching, to help calm your mind and body before sleep.

THE MOONLIT SANCTUARY: SLEEP HYGIENE AND CREATING A SLEEP-

FRIENDLY ENVIRONMENT

Crafting a sleep-friendly environment is like designing the moonlit sanctuary of our brain's city, providing the ideal conditions for rest and rejuvenation. Here are some tips for optimizing your sleep environment and promoting sleep hygiene:

Keep your bedroom cool, dark, and quiet: These conditions can help signal your brain that it's time to sleep and create a more comfortable environment for rest.

Invest in a comfortable mattress and pillows: Your sleep surface plays a crucial role in the quality of your rest, so invest in a mattress and pillows that provide the right support and comfort for your body.

Limit bedroom activities: Reserve your bedroom for sleep and intimacy only, avoiding activities such as work, television, or using your phone.

Remove distractions: Remove any sources of noise or light, such as electronic devices or ticking clocks, that may disturb your sleep.

Use soothing scents: Incorporate calming scents, such as lavender or chamomile, through essential oils or candles, to create a relaxing atmosphere conducive to sleep.

By mastering the art of slumber and crafting the moonlit sanctuary of your sleep environment, you will provide your brain with the essential rest and restoration it needs to thrive. As you prioritize restful, restorative sleep, you'll begin to experience the profound impact that quality sleep can have on your cognitive function, memory, and overall well-being.

Embrace the power of sleep in rejuvenating the cityscape of your mind and let the tranquil, moonlit sanctuary guide you towards a future filled with clarity, focus, and boundless potential. Tend to the streets and corners of your brain's city with the nurturing, regenerative force of sleep, and watch as it continues to sparkle and buzz with life, reflecting the vibrant, thriving landscape of your mental well-being.

CHAPTER 7: THE SERENE OASIS: EMBRACING MINDFULNESS AND MEDITATION FOR A FLOURISHING MIND

Picture your brain as an intricate labyrinth, an enchanted maze where thoughts, emotions, and memories intertwine and intersect, weaving the tapestry of your inner world. Within this complex network lies a serene oasis, a tranquil haven where mindfulness and meditation can cultivate clarity, focus, and inner peace. In this chapter, we will explore the profound benefits of mindfulness and meditation on brain health, delve into various practices, and learn how to develop a daily meditation routine.

THE HEALING WATERS: MINDFULNESS AND MEDITATION'S IMPACT ON BRAIN HEALTH

The healing waters of mindfulness and meditation gently flow through the labyrinth of our brain, nurturing the growth of new neurons and fostering synaptogenesis. Regular practice can enhance cognitive function, improve memory, and reduce stress and anxiety. By immersing ourselves in the serene oasis of meditation, we cultivate a fertile environment where our brain can flourish and thrive.

THE MANY PATHS TO TRANQUILITY: DIFFERENT TYPES OF MEDITATION AND MINDFULNESS PRACTICES

There are many paths leading to the serene oasis of meditation within our brain's labyrinth. Each route offers unique and transformative experiences that can foster brain health and well-being. Here are some meditation and mindfulness practices to explore:

Focused attention meditation: This practice involves concentrating on a single point, such as your breath, a mantra, or a visual object, to cultivate focus and stillness.

Open monitoring meditation: This form of meditation encourages non-judgmental awareness of your thoughts, feelings, and sensations as they arise, fostering a sense of acceptance and inner peace.

Loving-kindness meditation: Also known as "Metta" meditation, this practice involves sending loving and compassionate thoughts to yourself and others, nurturing empathy, and emotional well-being.

Body scan meditation: This practice involves bringing awareness to each part of your body, helping you cultivate a deeper connection to your physical sensations and promoting relaxation.

Mindful movement: Practices such as yoga, tai chi, and qigong combine movement, breath, and mindfulness to create a holistic mind-body experience.

CULTIVATING SERENITY: DEVELOPING

A DAILY MEDITATION ROUTINE

Creating a daily meditation routine is like cultivating the serene oasis within your brain's labyrinth, ensuring that the healing waters of mindfulness continue to flow and nourish your mind. Here are some tips for developing a daily meditation practice:

Start small: Begin with just a few minutes each day, gradually increasing the duration of your meditation sessions as you become more comfortable and experienced.

Choose a consistent time and place: Establish a designated time and location for your meditation practice, creating a tranquil and inviting space that encourages relaxation and focus.

Experiment with different techniques: Explore various meditation and mindfulness practices to find the one that resonates with you and supports your brain health goals.

Be patient and gentle with yourself: Remember that meditation is a journey, and it's normal to encounter challenges along the way. Approach your practice with patience, compassion, and curiosity.

Track your progress: Keep a meditation journal or use an app to monitor your practice and reflect on your experiences, making adjustments as needed to continue fostering brain health and well-being.

By embracing mindfulness and meditation, you will unearth the serene oasis within your brain's labyrinth, providing a tranquil haven for growth, healing, and self-discovery. Nurture your mind with the healing waters of meditation, and watch as your brain flourishes into a vibrant, thriving sanctuary of mental well-being.

As you continue to nurture the tranquil oasis of mindfulness and meditation within your brain's labyrinth, you will find that the benefits seep into every corner of your life. The clarity, focus, and inner peace cultivated through your practice will guide you in navigating the intricate maze of your thoughts, emotions, and memories, allowing you to make more mindful choices and foster a deeper connection to yourself and the world around you.

Embrace the transformative power of mindfulness and meditation in illuminating the pathways of your mind, and let the serene oasis be your compass as you journey through the labyrinth of your inner world. Immerse yourself in the healing waters of meditation, and watch as the tapestry of your thoughts, emotions, and memories begins to shimmer and sparkle with the vibrant colors of mental well-being.

As you delve deeper into your meditation practice, you may find new, undiscovered paths within your brain's labyrinth, leading to hidden treasures of insight, wisdom, and self-awareness. Follow these paths with curiosity and openness, allowing the transformative power of mindfulness and meditation to guide you on a journey of self-discovery, growth, and boundless potential.

In the end, the serene oasis of meditation becomes more than just a haven within your brain's labyrinth; it becomes a sanctuary within your heart, a space where you can always return to find solace, inspiration, and inner peace. As you continue to cultivate this sacred space through daily meditation practice, you'll find that the boundaries of the oasis expand, encompassing the entirety of your mental landscape, and transforming your brain into a thriving, flourishing garden of well-being.

CHAPTER 8: THE MIND'S PLAYGROUND: UNLOCKING POTENTIAL THROUGH COGNITIVE TRAINING AND BRAIN GAMES

I magine your brain as a vast, enchanted garden, where neurons bloom like flowers and synapses weave intricate pathways like ivy-covered trellises. Within this vibrant landscape lies a hidden playground, a whimsical space where cognitive training and brain games can spark curiosity, enhance mental agility, and promote neurogenesis and synaptogenesis. In this chapter, we will explore the role of cognitive training in fostering brain health, delve into popular brain games and exercises, and learn how to incorporate these engaging activities into your daily routine.

THE FERTILE GROUND: COGNITIVE TRAINING'S IMPACT ON NEUROGENESIS AND SYNAPTOGENESIS

Cognitive training and brain games serve as the fertile ground for our brain's enchanted garden, providing the essential nutrients and stimulation needed to promote the growth of new neurons and the formation of synapses. By engaging in regular cognitive training, we can enhance our mental agility, improve memory, and foster greater creativity and problem-solving skills. As we nurture our brain's garden with the stimulating activities of cognitive training, we can watch it flourish into a more resilient, adaptable, and vibrant landscape.

THE WHIMSICAL ATTRACTIONS: POPULAR BRAIN GAMES AND COGNITIVE EXERCISES

Within the mind's playground, a myriad of whimsical attractions awaits, each offering unique and engaging challenges designed to cultivate cognitive growth and mental agility. Here are some popular brain games and cognitive exercises to explore:

Sudoku: This number-based puzzle game challenges your logic and problem-solving skills, while also stimulating the growth of new neurons.

Crossword puzzles: These word-based puzzles help improve your vocabulary, memory, and language skills.

Memory games: From classic card matching games to more advanced exercises, memory games can sharpen your recall and enhance working memory.

Brain training apps: Many apps, such as Lumosity and Elevate, offer a wide range of cognitive exercises designed to target specific areas of brain function, such as memory, attention, and problem-solving.

Creative activities: Engaging in artistic pursuits, such as painting, drawing, or writing, can help stimulate your imagination and enhance your brain's overall cognitive flexibility.

CULTIVATING A MINDFUL ROUTINE: TIPS FOR INCORPORATING COGNITIVE TRAINING INTO YOUR DAILY LIFE

Incorporating cognitive training and brain games into your daily routine is like tending to the enchanted garden of your mind, ensuring that each neuron flower continues to blossom, and each synaptic pathway remains strong and vibrant. Here are some tips for integrating cognitive training into your daily life:

Set aside dedicated time: Carve out a specific time each day for cognitive training, whether it's during your morning coffee, your lunch break, or as part of your evening wind-down routine.

Choose activities that you enjoy: Select brain games and cognitive exercises that you find enjoyable and engaging, as you'll be more likely to stick with them in the long run.

Mix it up: Rotate between different types of cognitive activities to keep your brain challenged and stimulated, as variety can help promote neurogenesis and synaptogenesis.

Challenge yourself: As you become more adept at a particular brain game or exercise, increase the difficulty level to continue fostering cognitive growth and improvement.

Share the experience: Engage in cognitive training with friends or family members, as social interaction can enhance the benefits and enjoyment of these activities.

By embracing cognitive training and brain games, you can unlock the full potential of your mind's playground, nurturing the enchanted garden of your brain and guiding it towards a future.

filled with curiosity, mental agility, and boundless potential. As you tend to the vibrant landscape of your mind, cultivating the fertile ground of cognitive training, you'll witness the garden flourish into a thriving, resilient sanctuary of mental well-being.

Allow the whimsical attractions of brain games and cognitive exercises to inspire you on your journey of self-discovery and growth. Let each challenge become an opportunity to explore the hidden depths of your brain's enchanted garden, unveiling new pathways and unlocking hidden treasures of insight, creativity, and intellectual prowess.

As you weave cognitive training into the tapestry of your daily life, you will foster an environment in which your brain can continue to evolve, adapt, and thrive. Embrace the transformative power of cognitive training and brain games in shaping the landscape of your mind and let the enchanted garden of your brain blossom into a vibrant, dynamic oasis of mental health and well-being.

In the end, the mind's playground becomes more than just a space for cognitive training and brain games; it becomes a haven for exploration, discovery, and growth. As you continue to nurture this sacred space through daily cognitive practice, you'll find that the boundaries of the playground expand, encompassing the entirety of your mental landscape, and transforming your brain into a thriving, flourishing sanctuary of cognitive vitality and intellectual wonder.

CHAPTER 9: THE RESILIENT FOREST: CULTIVATING STRENGTH AND ADAPTABILITY TO OVERCOME LIFE'S OBSTACLES

E nvision your brain as an ancient, mystical forest, where the roots of your thoughts, emotions, and experiences intertwine to form a complex and ever-changing landscape. Within this enchanted realm, resilience acts as the life force that sustains the forest, allowing it to weather storms and adapt to new challenges. In this chapter, we will explore the importance of resilience in achieving your dreams, delve into strategies for cultivating resilience and overcoming obstacles, and learn how to harness setbacks as opportunities for growth and transformation.

THE SUSTAINING LIFE FORCE: THE IMPORTANCE OF RESILIENCE IN ACHIEVING YOUR DREAMS

Resilience is the life force that flows through the enchanted forest of your mind, nurturing the trees of your dreams and aspirations, and providing them with the strength to withstand adversity and flourish despite challenges. By cultivating resilience, you empower yourself to persevere in the face of setbacks, adapt to change, and ultimately, achieve your dreams. As you nurture your brain's resilient forest, you create a solid foundation upon which to build a life of purpose, passion, and boundless potential.

THE ROOTS OF RESILIENCE: STRATEGIES FOR DEVELOPING RESILIENCE AND OVERCOMING CHALLENGES

To cultivate the enchanted forest of resilience within your brain, you must nurture its roots with the nourishing waters of self-compassion, adaptability, and determination. Here are some strategies for developing resilience and overcoming life's obstacles:

Embrace self-compassion: Treat yourself with kindness and understanding, acknowledging your struggles, and offering yourself the same support and encouragement that you would extend to a dear friend.

Cultivate a growth mindset: View challenges as opportunities to learn and grow, recognizing that setbacks and failures are a natural part of the journey toward success.

Develop problem-solving skills: Approach obstacles with curiosity and creativity, brainstorming potential solutions and taking proactive steps to address the challenges you face.

Foster strong social connections: Build a supportive network of friends, family, and mentors who can offer guidance, encouragement, and a listening ear when you face difficulties.

Practice mindfulness and meditation: Engage in regular mindfulness and meditation practices to enhance emotional regulation, self-awareness, and stress management, all of which contribute to greater resilience.

THE SILVER LINING: HARNESSING SETBACKS AS OPPORTUNITIES FOR GROWTH

In the enchanted forest of your mind, even the darkest storms can bring forth a silver lining, offering opportunities for growth, renewal, and transformation. To harness setbacks as opportunities for growth, consider the following steps:

Reflect on your experiences: Take time to examine the setbacks and challenges you encounter, seeking to understand the lessons and insights they may hold.

Embrace adaptability: Be open to change and willing to adjust your course when necessary, recognizing that the path to success may not always be linear or predictable.

Celebrate small victories: Acknowledge and celebrate your accomplishments, no matter how small or seemingly insignificant, as each victory contributes to your overall growth and resilience.

Learn from failure: Instead of dwelling on setbacks, view them as valuable learning experiences, using the insights gained to inform future decisions and actions.

Practice gratitude: Cultivate an attitude of gratitude, focusing on the positive aspects of your life and the lessons you can learn from each challenge you face.

As you nurture the enchanted forest of resilience within your mind, you will find that the storms of life become less daunting, and the path to your dreams grows clearer and more accessible. By embracing self-compassion, adaptability, and determination, you can weather life's storms and transform setbacks into opportunities for growth and self-discovery.

Allow the resilience of the enchanted forest to be your guiding force as you traverse the complex landscape of your thoughts, emotions, and experiences. Let the nourishing waters of self-compassion and determination strengthen the roots of your dreams, enabling them to withstand adversity and flourish in the face of challenges.

As you cultivate resilience within the enchanted forest of your mind, you will find that the shadows of setbacks and obstacles give way to the light of growth, transformation, and boundless potential. Embrace the transformative power of resilience in shaping the landscape of your mind and let the enchanted forest of your brain become a sanctuary of strength, adaptability, and unwavering determination.

In the end, the enchanted forest of resilience becomes more than just a bastion of strength within your mind; it becomes a symbol of your unyielding spirit, your capacity to overcome adversity, and your unwavering commitment to achieving your dreams. As you continue to cultivate the resilient forest within your mind, you'll find that the boundaries of the forest expand, encompassing the entirety of your mental landscape, and transforming your brain into a thriving, flourishing sanctuary of fortitude and perseverance.

CHAPTER 10: THE SYMPHONY OF DREAMS: ORCHESTRATING YOUR LIFE FOR UNBOUNDED GROWTH AND FULFILLMENT

I magine your brain as a grand symphony hall, where the harmonious notes of neurogenesis, synaptogenesis, and personal growth resound in an exquisite, ever-evolving melody. In this final chapter, we will explore how to compose the symphony of your dreams by creating a personalized plan for harnessing neurogenesis and synaptogenesis, setting, and achieving goals, staying motivated and tracking progress, and embracing a growth mindset and lifelong learning.

COMPOSING THE SYMPHONY: CREATING A PERSONALIZED PLAN FOR HARNESSING NEUROGENESIS AND SYNAPTOGENESIS

To compose the symphony of your dreams, you must first craft a personalized plan that incorporates the various elements of neurogenesis, synaptogenesis, and personal growth. Consider the following steps:

Assess your current situation: Take an honest inventory of your current lifestyle, identifying areas that may be hindering your brain health and personal growth.

Set clear, achievable goals: Establish realistic and measurable goals in each area of your life, such as career, relationships, health, and personal development.

Develop a comprehensive strategy: Create a detailed plan that incorporates the strategies discussed in previous chapters, such as nutrition, exercise, sleep, mindfulness, cognitive training, and resilience.

Establish a timeline: Set a timeframe for achieving your goals, and break them down into smaller, manageable milestones.

Seek support: Engage friends, family, or a professional coach or mentor to provide encouragement, guidance, and accountability as you work toward your goals.

THE CRESCENDO OF SUCCESS: SETTING AND ACHIEVING GOALS

As the conductor of your life's symphony, setting and achieving goals is akin to orchestrating the crescendos and climaxes that define the masterpiece of your dreams. To bring your goals to fruition:

Prioritize your goals: Identify the most important goals and focus on achieving them first.

Break down goals into manageable steps: Divide your goals into smaller, achievable tasks, and tackle them one at a time.

Stay consistent: Dedicate time each day to work on your goals, even if it's just a few minutes.

Monitor your progress: Regularly assess your progress and adjust your plan as needed to stay on track.

Celebrate your accomplishments: Acknowledge and reward yourself for each milestone achieved, no matter how small.

THE RHYTHM OF PROGRESS: STAYING MOTIVATED AND TRACKING PROGRESS

Maintaining motivation and tracking progress is essential to ensuring that the symphony of your dreams continues to play in perfect harmony. To stay motivated and monitor your progress:

Keep a journal: Record your daily activities, thoughts, and achievements, using it as a tool to reflect on your progress and stay focused on your goals.

Set intermediate milestones: Establish smaller, short-term goals to work toward as you progress toward your larger, long-term goals.

Visualize success: Regularly visualize yourself achieving your goals and use this imagery to fuel your motivation and persistence.

Stay accountable: Share your goals and progress with a trusted friend or mentor who can provide support and encouragement.

Learn from setbacks: When faced with obstacles or setbacks, reflect on the experience, and use it as an opportunity for growth and improvement.

THE ETERNAL MELODY: EMBRACING A GROWTH MINDSET AND LIFELONG LEARNING

To ensure that the symphony of your dreams continues to evolve and resonate throughout your life, embrace a growth mindset, and commit to lifelong learning. By doing so, you'll foster an insatiable curiosity and a boundless thirst for knowledge, ensuring that your brain's symphony remains fresh, dynamic, and ever-expanding.

Stay curious: Cultivate a sense of wonder and curiosity about the world around you, seeking out new experiences, ideas, and perspectives.

Embrace challenges: View obstacles and setbacks as opportunities for growth, recognizing that the process of overcoming adversity contributes to your personal and intellectual development.

Seek out new learning opportunities: Pursue new hobbies, attend workshops, read widely, and engage in intellectual discussions to expand your knowledge and skills.

Surround yourself with like-minded individuals: Build a community of friends and mentors who share your passion for growth, learning, and personal development.

Reflect on your progress: Regularly evaluate your personal and intellectual growth, identifying areas for improvement and celebrating your successes.

As you compose the symphony of your dreams, you become the maestro of your own life, orchestrating a masterpiece of personal growth, self-discovery, and boundless potential. By creating a personalized plan for harnessing neurogenesis and synaptogenesis, setting, and achieving goals, staying motivated and tracking progress, and embracing a growth mindset and lifelong learning, you will transform your brain into a vibrant symphony hall, resounding with the harmonious melodies of success, happiness, and fulfillment.

In the end, the symphony of your dreams becomes more than just a celebration of your personal growth and achievement; it becomes the soundtrack of your life, a testament to your unwavering commitment to self-discovery, growth, and the pursuit of your highest potential. As you continue to compose the ever-evolving masterpiece of your life's symphony, you'll find that the boundaries of the symphony hall expand, encompassing the entirety of your mental landscape, and transforming your brain into a thriving, flourishing sanctuary of boundless creativity, curiosity, and passion.

CHAPTER 11: A JOURNEY THROUGH THE STARS: EMBRACING THE INFINITE POTENTIAL OF NEUROGENESIS AND SYNAPTOGENESIS

As we reach the conclusion of our voyage through the cosmos of the mind, let us pause to reflect on the celestial wonders we have encountered, and the limitless potential for transformation that awaits us through the power of neurogenesis and synaptogenesis. Like intrepid explorers charting new territories among the stars, we have discovered that the vast expanse of our inner universe holds infinite potential for growth, evolution, and personal fulfillment.

THE INFINITE GALAXY: THE POTENTIAL FOR TRANSFORMATION THROUGH NEUROGENESIS AND SYNAPTOGENESIS

Throughout our journey, we have witnessed the awe-inspiring power of neurogenesis and synaptogenesis to reshape the very fabric of our brains, opening the doors to new possibilities and uncharted realms of personal growth. Like the birth of a supernova, these processes illuminate our minds with the radiant energy of new neurons and synaptic connections, empowering us to transcend our limitations and reach for the stars.

Just as the celestial bodies within the galaxy are in constant motion, our brains are ever-evolving landscapes of neural activity, constantly adapting and transforming in response to our thoughts, emotions, and experiences. By harnessing the power of neurogenesis and synaptogenesis, we can become the architects of our own minds, sculpting our neural pathways and synaptic connections to create a vibrant, flourishing universe within our brains.

EMBARKING ON THE ODYSSEY: ENCOURAGEMENT TO EMBARK ON THE JOURNEY TO LIVING THE LIFE OF YOUR DREAMS

Now, as you stand on the precipice of a new frontier, gazing into the vast expanse of your own potential, we invite you to embark on the greatest odyssey of your life – the journey to living the life of your dreams. Like a celestial explorer venturing forth into the unknown, you have the power to chart your own course, navigate the boundless expanses of your mind, and forge a path to the stars.

Armed with the tools and strategies we have explored throughout this book, you possess the keys to unlocking the infinite potential of your brain and tapping into the limitless reservoir of creativity, innovation, and resilience that resides within you. As you embrace the power of neurogenesis and synaptogenesis, you will find that the boundaries of your mental universe expand, giving way to a dazzling panorama of personal growth, self-discovery, and boundless potential.

As you embark on your celestial odyssey, we encourage you to stay true to your course, persevere in the face of adversity, and never lose sight of the shimmering constellations that guide your way. For within the cosmic tapestry of your mind lies an infinite universe of possibilities, waiting to be explored, charted, and illuminated by the radiant energy of your dreams.

In the end, the journey through the stars becomes more than just an odyssey of personal growth and self-discovery; it becomes a testament to the boundless potential of the human spirit, an enduring reminder of our capacity to transcend our limitations and soar to the farthest reaches of our dreams. As you continue to navigate the celestial realms of your mind, let the infinite potential of neurogenesis and synaptogenesis become the guiding light that illuminates your path, leading you ever onward, upward, and beyond.

EPILOGUE: A BEACON OF LIGHT IN THE COSMIC EXPANSE

As we bring our celestial odyssey to a close, let us not view this as the end of our journey, but rather as the beginning of a new chapter in the ever-unfolding saga of our lives. Our voyage through the cosmos of the mind has illuminated the infinite potential that resides within each of us, and we must now carry that light forward, like a beacon of hope and inspiration, guiding us as we venture forth into the uncharted realms of our dreams.

As you continue to explore the boundless expanse of your inner universe, remember that the power of neurogenesis and synaptogenesis resides within you, and it is up to you to harness that energy and use it to transform your life. Embrace the potential for growth, change, and self-discovery, and let the radiant glow of your dreams illuminate the path before you, guiding you towards a brighter, more fulfilling future.

And as you venture forth into the unknown, remember that you are not alone. Like the constellations that adorn the night sky, we are all connected by a shared tapestry of dreams, aspirations, and desires, each of us on our own celestial journey, navigating the vast expanse of our inner cosmos. As you continue to chart your course among the stars, know that you are part of a greater community of explorers, each of us striving to unlock the limitless potential that resides within our hearts and minds.

Together, we can create a universe of infinite possibility, a celestial symphony of growth, learning, and self-discovery that resounds throughout the cosmos, echoing the timeless wisdom of the ages: that within each of us lies the power to shape our destiny, forge our own path, and reach for the stars.

So, as you stand on the precipice of your own cosmic adventure, take a moment to gaze up at the night sky, and let the infinite expanse of the universe serve as a reminder of the boundless potential that lies within you. Remember that, like the stars that shimmer in the celestial tapestry, your dreams are the guiding light that will lead you to your destiny, and the radiant energy of neurogenesis and synaptogenesis is the fuel that will propel you on your journey to the farthest reaches of your dreams.

With this newfound knowledge and the tools, you have acquired, set forth with courage, determination, and an unquenchable thirst for discovery. Embrace the cosmic adventure that lies before you and let the infinite potential of your mind be the compass that guides you on your journey through the stars.

And as you embark on this grand odyssey, remember the words of the great poet Rumi: "You are the entire ocean in a drop." Within you lies an infinite universe of possibility, a cosmic expanse of dreams and potential that is waiting to be explored, charted, and illuminated by the radiant energy of your soul. So go forth, dear explorer, and let the symphony of your dreams resound throughout the cosmos, a testament to the boundless potential of the human spirit, and the unquenchable thirst for growth, learning, and self-discovery that lies at the heart of our existence.

Bon voyage, intrepid traveler, and may the stars forever light your way.

POSTSCRIPT: THE ETERNAL VOYAGE OF SELF-DISCOVERY

As our celestial journey comes full circle, we find ourselves gazing back at the path we have traveled, reflecting on the myriad lessons, experiences, and insights we have gained along the way. The voyage through the cosmos of our minds has been one of transformation, growth, and self-discovery, and as we look to the horizon, we find ourselves at the dawn of a new era, poised to embark on a never-ending odyssey of personal exploration and enlightenment.

For the journey of self-discovery is an eternal one, a voyage that knows no end, as we continually strive to unlock the boundless potential that resides within our hearts and minds. Each new experience, each triumph and setback, serves as a steppingstone on our path to self-mastery, propelling us ever onward, upward, and beyond the limits of our dreams.

As you continue to navigate the vast expanse of your inner cosmos, remember that you are the master of your own destiny, the architect of your own dreams, and the captain of your own celestial ship. The power of neurogenesis and synaptogenesis is yours to wield, a potent force for transformation, growth, and self-discovery that lies dormant within the depths of your being.

As you venture forth into the unknown, embrace the spirit of adventure, curiosity, and wonder that has guided you on your celestial journey thus far. Seek out new horizons, forge new connections, and continue to cultivate the boundless potential that lies within your heart and mind.

And as you chart your course among the stars, remember that the journey of self-discovery is a shared one, a cosmic dance of dreams, aspirations, and desires that connects us all, uniting us in our quest for knowledge, growth, and personal fulfillment. As you continue to explore the celestial realms of your inner universe, know that you are part of a greater tapestry of human potential, a constellation of dreamers and visionaries who, together, are forging a brave new world of infinite possibility, boundless creativity, and uncharted realms of personal growth.

So, as we bid farewell to our cosmic adventure, let us not view this as the end of our journey, but rather as the beginning of a new chapter in the ever-evolving saga of our lives. May the spirit of discovery, wonder, and boundless potential that has guided us on our celestial voyage continue to illuminate our path, guiding us ever onward, upward, and beyond the farthest reaches of our dreams.

And as you embark on the eternal voyage of self-discovery, may the stars forever light your way, and the boundless expanse of the cosmos serve as a reminder of the infinite potential that lies within us all.

Farewell, intrepid traveler, and may the journey of a lifetime continue to unfold before your eyes, an enduring testament to the boundless potential of the human spirit, and the unquenchable thirst for growth, learning, and self-discovery that lies at the heart of our existence.

www.ingramcontent.com/pod-product-compliance
Lightning Source LLC
Chambersburg PA
CBHW070517220526
45467CB00002B/708